Daniel Dantas
Maria José Hatem de Souza

Analyses of longitudinal climate data in cities in Minas Gerais

Daniel Dantas
Maria José Hatem de Souza

Analyses of longitudinal climate data in cities in Minas Gerais

Climate modelling and temporal variations in precipitation, rainy periods, rainfall erosivity and veraison

ScienciaScripts

Imprint

Cover image: www.ingimage.com

This book is a translation from the original published under ISBN 978-613-9-66415-3.

Publisher:
Sciencia Scripts
is a trademark of
Dodo Books Indian Ocean Ltd. and OmniScriptum S.R.L publishing group

120 High Road, East Finchley, London, N2 9ED, United Kingdom
Str. Armeneasca 28/1, office 1, Chisinau MD-2012, Republic of Moldova, Europe
Printed at: see last page
ISBN: 978-620-8-02741-4

SUMMARY

CHAPTER 1

EROSIVE POTENTIAL OF RAINFALL IN FIVE CITIES IN MINAS GERAIS, FROM 1961 TO 2015

Daniel Dantas[1] , Maria José Hatem de Souza[2] , Ricardo Andrade Generoso[3]

[1] Master's student in Forestry Engineering, UFLA, Lavras, Minas Gerais, dantasdaniel12@yahoo.com.br;[2] Agricultural Engineer, Associate Professor, Agronomy Department, UFVJM, Diamantina-MG;[3] Undergraduate student in Science and Technology; UFVJM, Diamantina-MG.

ABSTRACT: The aim of this study was to estimate the average monthly (Ei30) and annual (R factor) erosivity and compare the variation of these factors in the periods 1961 to 1990 and 1991 to 2015 for the municipalities of Aimorés, Araxá, Bambuí, Barbacena and Belo Horizonte in Minas Gerais. Monthly rainfall data from historical rainfall series belonging to the National Meteorological Institute (INMET) and the Hydrological Information System of the National Water Agency (ANA) were used to assess the variation in erosivity over time. To determine the average monthly erosivity index, the average monthly rainfall for each period was calculated. The average annual rainfall erosivity index (R factor) was obtained by adding up the monthly EI30 values. The erosivity index (EI30) is closely related to the amount of rainfall, with the months of January and December having the highest average rainfall values and, consequently, the highest average monthly rainfall erosivity values. The average annual erosivity values found indicated high erosivity for the cities of Araxá, Bambuí and Belo Horizonte in both periods. BH was the city with the highest R factor values, with 8469 and 9227 MJ.mm/ha.h.year in periods 1 and 2, respectively. The city of Aimorés showed a moderate annual erosivity index, while Barbacena showed an increase in erosivity from moderate to high between the two periods.

KEY WORDS: erosivity, precipitation, water erosion

INTRODUCTION

Soil degradation is mainly caused by the dragging away of smaller, more nutrient-rich particles, culminating in a decrease in fertility and, consequently, a reduction in yields or an increased need for fertilisers and correctives. In most cases, soil losses caused by water erosion reduce the thickness of the soil, decreasing the capacity to retain and redistribute water in the profile, generating, as a consequence, greater surface runoff and, sometimes, higher rates of soil erosion (SANTOS et al., 2010). In short, water erosion has a high potential for reducing the productive capacity of soils and can jeopardise surface water resources (MELLO et al., 2007).

The association of characteristics such as disturbed terrain, inadequate land use and management and adverse physical and water characteristics with rainfall of greater intensity and frequency increases the risk of erosion (SANTOS et al., 2010). In the erosion process, among the various factors related to the occurrence of erosion, rainfall erosivity is one of the most important (MELLO et al., 2007). The R factor, called rainfall erosivity, which makes up the universal equation for soil loss, is a numerical index that expresses the capacity of rainfall, expected in a given location, to cause water erosion in an unprotected area (BERTONI & LOMBARDI NETO, 1993). It can be expressed using indices based on the physical characteristics of rainfall in each region (CABRAL et al., 2005).

Determining erosivity values makes it possible to identify the months in which the risk of soil and water loss is highest, which is why it plays an important role in planning conservation practices based on maximum soil cover in the critical seasons when rainfall has the greatest erosive capacity (WISCHMEIER & SMITH, 1978; BERTONI & LOMBARDI NETO, 1993; HUDSON, 1995). Many studies have been carried out in order to define the spatial distribution of rainfall erosivity in regions of Minas Gerais, such as Aquino (2005), for the southern region of Minas Gerais; and Mello et al. (2007), for the state of Minas Gerais.

In view of the above, the aim was to estimate the average monthly (EI30) and annual (R factor) erosivity and compare the variation of these factors in the time series from 1961 to 2015 in five cities in Minas Gerais.

MATERIAL AND METHODS

Monthly rainfall data from historical rainfall series obtained from the National Meteorological Institute (INMET) and the Hydrological Information System of the National Water Agency (ANA) were used. The five cities analysed are located in the state of Minas Gerais: Aimorés, belonging to the Vale do Rio Doce mesoregion and Aimorés micro-region; Araxá, located in the Triângulo Mineiro and Alto Paranaíba mesoregion; Bambuí, located in the centre-west of Minas Gerais, near the Serra da Canastra; Barbacena, located in the Campo das Vertentes mesoregion; and Belo Horizonte, in the Belo Horizonte metropolitan mesoregion.

Information on the meteorological stations in the respective cities considered in this study is shown in Table 1.

Table 1. INMET weather stations involved in this study, World Meteorological Organisation (WMO) number, geographical coordinates, operating status and when opened, and number of years involved in the study.

Station	**Latitude** (degrees)	**Longitude** (degrees)	**Altitude** (metres)
Aimorés (OMM: 83595)	-19,49	-41,07	82,74
Araxá (OMM: 83579)	-19,6	-46,94	1023,61
Bambuí (OMM: 83582)	-20,03	-45	661,27
Barbacena (OMM: 83689)	-21,25	-43,76	1126
Belo Horizonte (OMM: 83597)	19,93	43,93	915

The data used in this study refers to the years 1961 to 2015, divided into two periods: 1961 to 1990 and 1991 to 2015, in order to analyse the behaviour of precipitation and, consequently, rainfall erosivity over the years.

The two-sided Student's t-test was applied to test whether or not there was a significant difference between the average annual rainfall volumes for periods 1 and 2.

To determine the erosive potential of rainfall, indices are used, such as the standard erosivity index, called EI30, described by Wischmeier & Smith (1978), which represents the product of the kinetic energy with which the raindrop hits the ground and its maximum intensity. According to Bertoni & Lombardi Neto (1993), this product represents an interaction term that measures the effect of how erosion by impact, splash and turbulence combines with the downpour to transport the detached soil particles.

To determine the average monthly erosivity index (EI30), the monthly average for each of the five periods evaluated was calculated using equation 1, proposed by Wischmeier & Smith (1978):

$$EI_{30} = 67{,}355\left(\frac{r^2}{P}\right)^{0{,}85} \quad (1)$$

Being:

r: average monthly rainfall (mm)

P: average annual rainfall (mm)

The annual rainfall erosivity index (R) is the sum of the monthly values of this index, according to equation 2:

$$R = \sum_{1}^{12} EI_{30} \quad (2)$$

The values found for the EI30 and the R factor were compared in order to check for changes in the time series studied. The two-sided Student's t-test was applied to test whether or not there was a significant difference between the monthly and annual rainfall erosivity indices for periods 1 and 2 for each city analysed in this study.

RESULTS AND DISCUSSION

Of the five cities analysed, three - Araxá, Bambuí and Belo Horizonte - showed an increase in average rainfall from period 1 (1961 to 1990) to period 2 (1991 to 2015) (Table 1). Belo Horizonte was the city with the greatest variation in rainfall between the two periods: in period 1 the annual average rainfall was 1480 mm, while in period 2 this figure was 1608 mm, an increase of 8.6%. In Bambuí, the average annual rainfall increased from 1398 mm to 1481 mm, an increase of 5.9%. And Araxá showed a variation from 1500 mm in average annual rainfall in period 1 to 1558 mm in period 2, an increase of 3.9%.

Table 1. Average monthly rainfall and R factor for each period evaluated

City	**Average annual rainfall (mm)**	
	1961-1990	**1991-2015**
Aimorés	1007	979
Araxá	1500	1558
Bambuí	1398	1481
Barbacena	1365	1267
Belo Horizonte	1480	1608

As for the cities that showed a reduction in average annual rainfall, in Aimorés the reduction was from 1007 mm in period 1 to 979 mm in period 2, a percentage difference of 2.8 per cent; and for Barbacena the reduction was from 1365 mm in period 1 to 1267 mm in period 2, 7.2 per cent.

Statistically, there was no significant difference between the annual rainfall volumes between periods 1 and 2 for any city, according to the student's t-test, carried out at a 95% confidence level. This indicates that although there were differences between the annual rainfall averages, these differences were not significant and it is not possible to say that there is a tendency for annual rainfall volumes to increase or decrease over the period studied for the five cities. It is worth noting, however, that this study analyses the volume of rainfall throughout the year, and not the dynamics of rainfall throughout the year and its isolated events.

The cities considered in this study have a well-defined dry and rainy season in which the driest quarter comprises the months of June, July and August and the quarter with the highest rainfall runs from November to January. The months of January and December stand out as having the highest average rainfall values

The behaviour of the average monthly rainfall (mm) for the five cities analysed, over the entire time series studied, for periods 1 and 2, is shown in Figure 1.

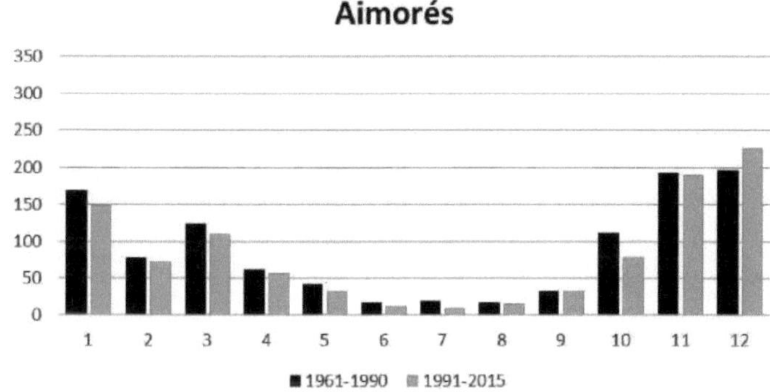

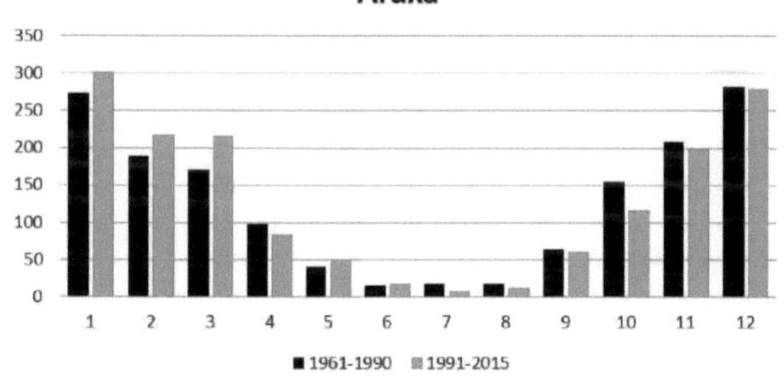

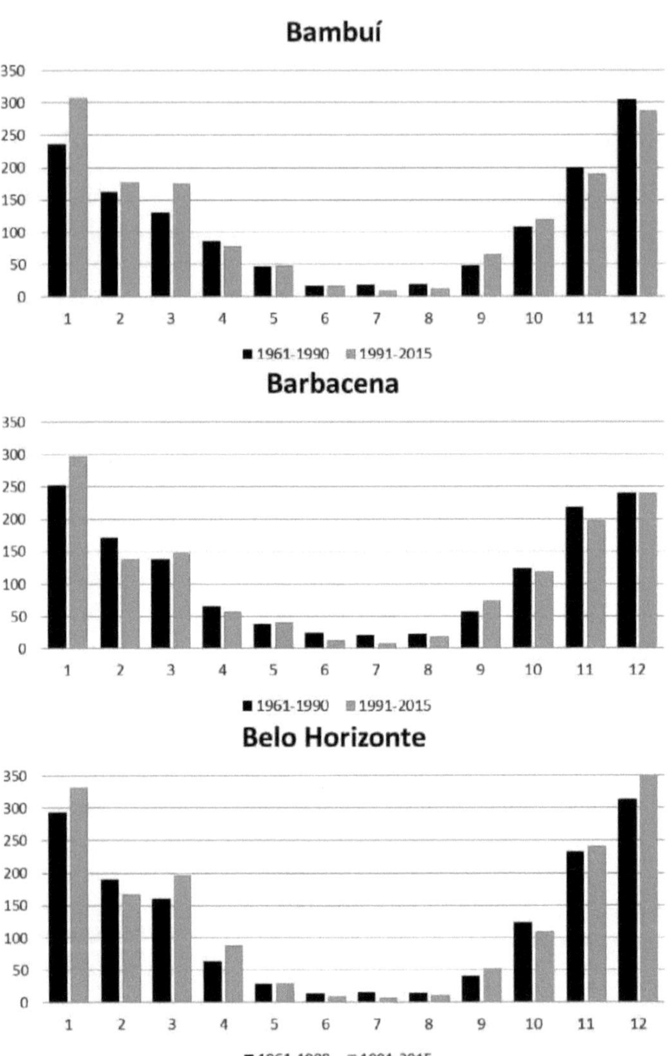

Figura 1. Average monthly rainfall (mm) during the periods studied for the cities of Aimorés, Araxá, Bambuí, Barbacena and Belo Horizonte in Minas Gerais.

It can be seen from the figures above that, in general, the average monthly rainfall volumes remained very close in all the months in both periods analysed. In Aimorés, it is worth highlighting the reduction in the average volume of precipitation in October from 111 mm in period 1 to 79 mm, while in December there was an increase in this volume, which went from 196 mm to 225 mm. In the cities of Araxá and Bambuí, there was an increase in average monthly rainfall in January, February and March, which influenced the increase in average annual rainfall in these cities.

For the city of Barbacena, there was an increase in average monthly rainfall in the month of January, from 252 mm in period 1 to 297 mm in period 2, and a reduction in average rainfall in 9 months of the year, with the exception of January, March and September. In Belo Horizonte, on the other hand, there was an increase in the average monthly rainfall in 7 months of the year, the highest being in December, where the values were 312 mm in period 1 and 356 mm in period 2.

The annual behaviour of the average monthly erosivity index (MJ.mm/ha.h) for the five cities over the entire time series studied is directly related to the variation in rainfall over the months of the year (Figure 1).

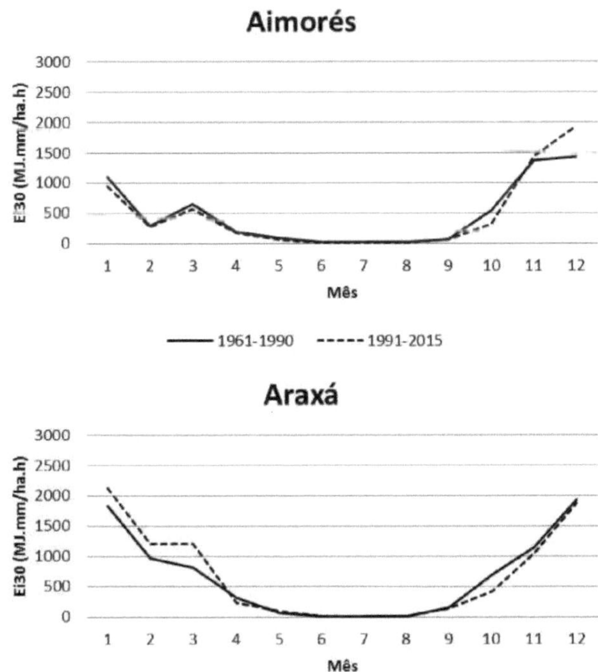

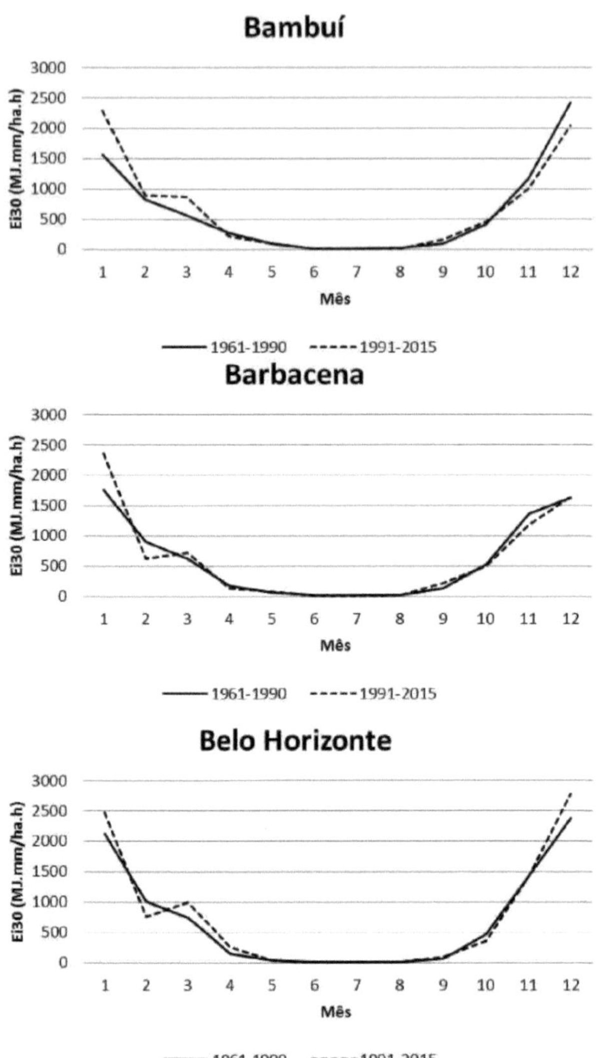

Figura 2. Behaviour of the average monthly erosivity index - Ei30 (MJ.mm/ha.h) in the periods studied, for the cities of Aimorés, Araxá, Bambuí, Barbacena and Belo Horizonte, in Minas Gerais.

In all the cities, there were practically zero erosivity indices between the months of May and August, with an increase in Ei30 from September onwards, with a decrease from January until April, when the rainy season ends.

In the city of Aimorés, the highest Ei30 value was found in December in both periods, with 1434 MJ.mm/ha.h in period 1 and 1930 MJ.mm/ha.h in period 2. In Araxá, the highest Ei30 value was found in December in period 1, with a value of 1936 MJ.mm/ha.h, and in January in period 2, with a value of 2133 MJ.mm/ha.h. In Bambuí, the same thing happened: in period 1, the average erosivity index was highest in December, while in period 2, the highest value was seen in January, at 2295 MJ.mm/ha.h. For the city of Barbacena, the average monthly erosivity indices remained at their highest levels in January, with 1766 MJ.mm/ha.h in period 1 and 2364 MJ.mm/ha.h in period 2. Belo Horizonte had the highest monthly erosivity index in December in both periods, with 2371 MJ.mm/ha.h in period 1 and 2782 MJ.mm/ha.h in period 2.

The average annual erosivity values (R Factor) for the cities and periods under study are shown in Table 2.

Table 2. Average monthly rainfall and R factor for each period evaluated

City	R Factor (MJ.mm/ha.h.year)	
	1961-1990	**1991-2015**
Aimorés	5837	5854
Araxá	8012	8418
Bambuí	7507	8105
Barbacena	7272	7500
Belo Horizonte	8469	9227

Belo Horizonte had the highest annual rainfall erosion rates (R Factor) among the cities analysed, with 8469 MJ.mm/ha.h.year in period 1 and 9227 MJ.mm/ha.h.year in period 2. The lowest values were observed in the city of Aimorés, which remained very close in both periods, due to the small variation in average annual rainfall between the periods, with the R factor equal to 5837 MJ.mm/ha.h.year for period 1 and 5854 MJ.mm/ha.h.year for period 2.

Considering the parameters shown in Table 3, according to the classification by Foster et al. (1981), the annual rainfall erosivity index remained in the High erosivity class in Araxá, Bambuí and Belo Horizonte in both periods. In Aimorés, the R factor remained in the Moderate erosivity class in periods 1 and 2.

Table 3. Interpretation classes of the mean annual erosivity index (R)

Erosivity (MJ.mm/ha.h.year)	Erosivity classes
R≤2452	Low
2452 <R≤4905	Average
4905 < R ≤7357	Moderate
7357<R≤9810	High
R > 9810	Very high

Source: (CARVALHO, 2008; FOSTER et al., 1981).

In Barbacena there was a change in the erosivity class from Moderate to High, since the observed R-factor values were 7272 MJ.mm/ha.h.year in period 1 and 7500 MJ.mm/ha.h.year in period 2. Despite the change in classification, it is worth noting that there was no significant variation in the annual erosivity values for Barbacena, which can be statistically proven by the t-test, applied at a 95 % confidence level, between the two periods. The same was observed for the other cities, which statistically showed no significant variation (p-value > 0.05) in the annual erosivity indices between periods 1 and 2.

CONCLUSIONS

The months with the highest average rainfall in the cities of Aimorés, Araxá, Bambuí, Barbacena and Belo Horizonte in Minas Gerais are January and December and, consequently, these are also the months with the highest average erosivity index values. This indicates the need for greater attention to be paid to soil management practices aimed at reducing rainfall erosivity during these periods of higher rainfall.

Despite an increase in the average volume of rainfall in three cities and, consequently, an increase in the annual rainfall erosivity index, no statistically significant differences (p-value > 0.05) were observed in any city between the periods 1961 to 1990 and 1991 to 2015.

ACKNOWLEDGEMENTS

The Research Support Foundation of the State of Minas Gerais for its financial support for the third author's scientific initiation scholarship, which allowed the data to be processed. The National Meteorological Institute (INMET) for the data provided.

BIBLIOGRAPHICAL REFERENCES

AQUINO, R. F. Rainfall patterns and spatial variability of erosivity for the south of the State of Minas Gerais. Lavras, Federal University of Lavras, 2005. 98p. (Master's Thesis)

BERTONI, J. & LOMBARDI NETO, F. Soil conservation. São Paulo, Ícone, 1993. 355p.

CABRAL, J.B.P.; BECEGATO, V.A.; SCOPEL, I. & LOPES, R.M. Study of erosivity and spatialisation of data with geoprocessing techniques on the topographic map of Morrinhos-Goiás/Brazil for the period 1971 to 2000. GeoFocus, 5:1- 18, 2005.

FOSTER, G. R.; MCCOOL, D. K.; RENARD, K. G.; MOLDENHAUER, W. C. Conservation of the universal soil loss equation to SI metric units. Journal of Soil and Water Conservation, v.36, p. 355-359, 1981.

HUDSON, N. Soil conservation. 2.ed. Ithaca, Cornell University Press, 1971. 320p.

LUCIO, P. S.; ABREU, M. L.; TOSCANO, E. M. M. Characterisation of climatological series in Belo Horizonte - MG. PART I, II, III, IV:, X Brazilian Congress of Meteorology, Brasília - DF, 1998.

MELLO, C.R.; SÁ, M.A.C.; CURI, N.; MELLO, J.M. & VIOLA, M.R. Monthly and annual rainfall erosivity in the state of Minas Gerais. Pesquisa Agropecuária Brasileira,

Brasília, v.42, n.4, p.537-545, Apr. 2007.

MELLO, C. R. de & SILVA, A. M. da. Statistical modelling of monthly and annual precipitation and the dry season for the state of Minas Gerais. Brazilian Journal of Agricultural and Environmental Engineering. v.13, n.1, p68-74, 2009

PINHEIRO, M. M. G. & BAPTISTA, M. B. Análise regional de frequência e distribuição temporal das tempestadas na região metropolitana de Belo Horizonte - RMBH, Revista Brasileira de Recursos Hídricos 3(4): 73-88, 1998.

SANTOS, G. G.; GRIEBELER, N. P.; OLIVEIRA, L. F. C. de. Intense rainfall related to water erosion. Brazilian Journal of Agricultural and Environmental Engineering. v.14, n.2, p. 155-123, 2010.

SILVA, A. M. da; SILVA, M. L. N.; CURI, N.; LIMA, J. M. de; AVANZI, J. C.; FERREIRA, M. M. Losses of soil, water, nutrients and organic carbon in Cambissolo and Latossolo under natural rainfall. Pesquisa Agropecuária Brasileira. v.40, p. 1223-1230, 2005.

WISCHMEIER, W. H. & SMITH, D. D. Predicting rainfall erosion losses: a guide to conservation planning. Washington, USDA, 1978. 58p. (Agriculture Hand-Book, 537).

CHAPTER 2

VERANICO INCIDENCE IN MONTES CLAROS, MINAS GERAIS

Daniel Dantas[1] , Maria José Hatem de Souza[2] , Bruno de Oliveira Fernandes[3] , Fabricio Resende Aguiar[4] , Luiza Pereira Sánchez[5] .

[1]Master's student in Forestry Engineering, UFLA, Lavras, Minas Gerais, dantasdaniel12@yahoo.com.br;[2] Agricultural Engineer, Associate Professor, Agronomy Department, UFVJM, Diamantina-MG;[3] Undergraduate student in Agronomy, UFVJM, Diamantina-MG;[4] Undergraduate student in Agronomy, UFVJM, Diamantina-MG;[5] Undergraduate student in Agronomy, UFVJM, Diamantina-MG.

ABSTRACT: The incidence of dry days during the rainy season can jeopardise the development of plants in agricultural crops. This study was carried out in order to assess the dynamics of the occurrence of veranicos during the months of November, December, January and February, which comprise the rainy season, in the city of Montes Claros, in the north of the state of Minas Gerais. Daily rainfall data from 1962 to 2015 was used. The data was obtained from the National Meteorological Institute - INMET. The length of the veranico was determined for each month, taking into account the longest sequence of dry days in each month, where a dry day is one in which rainfall is equal to or less than 1 mm. The city of Montes Claros experiences long periods of summer drought. The average number of summer days for the city is 8.97 days in November, 9.73 days in December, 12.86 days in January and 12.29 days in February. The maximum veranico value was observed in December 2014, lasting 44 days. Analysing the veraison values, it would be recommended to start planting in November, for example, to grow crops with higher water requirements in the initial phase, as this month has, on average, the shortest sequence of dry days.

KEY WORDS: rainy season, dry days, northern Minas Gerais.

INTRODUCTION

According to Carvalho et al. 1999, rainfall in the state of Minas Gerais, like almost all Brazilian states, is not evenly distributed at all times of the year. There are rainy periods and dry periods. In the Southeast, the summer period is characterised by rainfall, the total amount of which normally satisfies the water needs of the major crops. However, rainfall data alone does not provide accurate information about the region's climate and whether the conditions are suitable for growing a particular crop under rainfed conditions. To do this, water storage parameters in the soil and moisture gains and losses in the soil-plant-atmosphere system must be analysed (MOTA, 1989). According to Neto and Vilela, 1986, it is still common for dry days to occur during the rainy season, a phenomenon known as veranico. Depending on the duration and stage of the crop in which the veranico occurs, final production can be seriously affected due to yield losses caused by water stress in the plants.

During the dry season and also during the long veranicos that occur in the middle of the rainy season, burn-offs in hills and scrubland are also common, especially in rural areas of the city, which contributes to deforestation and the release of pollutants into the atmosphere, further damaging air quality (Paula Machado, 2009). It is therefore important to carry out studies that can provide information on this phenomenon in order to help plan tree planting and crop management.

The aim of this study was to determine the cumulative probability of veranicos during the rainy season (November to February) for the city of Montes Claros, Minas Gerais.

MATERIAL AND METHODS

Montes Claros is a Brazilian municipality in the north of the state of Minas

Gerais. It belongs to the microregion of the same name and the Mesoregion of Norte de Minas. It covers an area of 3 582.034 km^2 , of which 38.7 km^2 are in the urban perimeter and the remaining 3 543.334 km^2 are in the countryside.

Montes Claros has a tropical climate, type Aw according to the Koppen classification, with less rainfall in winter and an average annual temperature of 22.4 °C, with dry, mild winters and rainy summers with high temperatures. The warmest month, February, has an average temperature of 23.8 °C, with an average maximum of 30.4 °C and a minimum of 19 °C. The coldest month, July, has an average temperature of 19.6 °C, with 27.4 °C and 12.5 °C being the maximum and minimum averages, respectively. Autumn and spring are transitional seasons.

The average rainfall is 1,086.4 millimetres (mm) per year, with July being the driest month, when only 0.5 mm occur. In December, the wettest month, the average is 230.9 mm. During the dry season and long summers in the middle of the rainy season, fires are also common on hillsides and in scrubland, especially in rural areas of the city, which contributes to deforestation and the release of pollutants into the atmosphere, further damaging air quality. The relative humidity is 67 per cent, and on some days of the year, especially in winter, it can drop below 30 per cent or even 20 per cent.

Daily rainfall data for Montes Claros was obtained from the National Meteorological Institute (INMET). Data was used for the period from January 1961 to July 2015. The Montes Claros A506 station is located at a latitude of -16.68°, a longitude of -43.84° and an altitude of 652 metres.

For each rainy season, the length of the veranicos was determined for the months of November to February. The longest sequence of dry days in each month was taken into account. A dry day is one in which rainfall was equal to or less than 1 mm, since values of less than 1 mm/day are difficult for crops to absorb and are quickly evaporated, thus maintaining the effect of the summer. When a continuous period of rain extends into the following month, the dry period will be calculated in the month of the day the dry period begins.

RESULTS AND DISCUSSION

For the city of Montes Claros, it is possible to see long periods of summer drought in the months analysed, considering that these months make up the region's rainy season, which runs from October to March.

Of the four months analysed, January is the month in which the longest veranico periods occur on average (Table 1). An average of 12.86 days of veraison were observed in January, followed by February, with an average of 12.29 days of veraison; December, with 9.73 days of veraison; and November, with 8.97 days.

Table 1. Average and maximum values of veranicos, in days, in Montes Claros, Minas Gerais, during the months of November, December, January and February.

Month	Average summers (days)	Maximum values (days)
November	8,97	24
December	9,73	44
January	12,86	35
February	12,19	38

The maximum values observed were in December 2014, when there was a veranico lasting 44 days. In November, the maximum value was 24 days in 1998. In January, the maximum veranico was 35 days in 2014. In February, the maximum veranico was 38 days in 2011. Assad et al. (1993) pointed out that the veranico is not atypical, but typical of the cerrado region. And that the north of Minas Gerais, where the city of Montes Claros is located, is one of the regions in Brazil with the highest frequency of veranicos longer than ten days in the months of January. Considering the 37 years analysed in this study, in 22 of them there were veranicos lasting more than ten days for the month of January. The same value was found for the month of

February. The variability of the rainfall recorded has caused damage to the agricultural sector and the region's economy. Although it rains as much as in many other regions of Brazil, the city of Montes Claros faces problems because of its spatio-temporal distribution and the magnitudes and intensities of its rainfall. Although it is possible to observe a slight upward trend in summer periods over the years, this trend has low r-squared values and shows high variability in the occurrence of data over the historical series analysed (Figure 1).

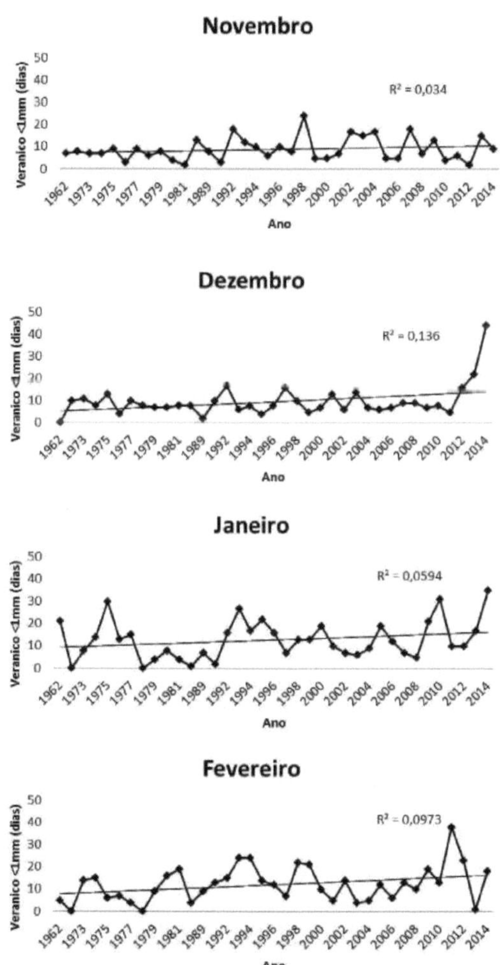

Figure 1. Variation over the years in the number of summer days in Montes Claros in

November, December, January and February since 1962.

Analysing the veranico values, which are restricted to the months of November, December, January and February, it would be recommended to start planting in November, for example, for crops with higher initial water requirements, as this month has, on average, the shortest sequence of dry days.

CONCLUSIONS

The city of Montes Claros faces long periods of summer, considering the historical series analysed for the months of November, December, January and February. The month of January has the longest summer periods. The average number of summer days for the city is 8.97 days in November, 9.73 days in December, 12.86 days in January and 12.29 days in February. Analysing the veranico values, it would be recommended to start planting in November, for example, to grow crops with higher water requirements in the initial phase, as this month has, on average, the shortest sequence of dry days.

REFERENCES

Assad, E. D., Sano, E. E., Masutomo, R., de Castro, L. H. R., & da Silva, F. A. M. (1993). Veranicos in the Brazilian cerrado region: frequency and probability of occurrence. *Pesquisa Agropecuária Brasileira,* 28(9), 993-1003.

Carvalho, D. F. Oliveira, M. A. A. Sousa, S. A. V. Carvalho, P. O. L. Estimates of the occurrence of veranicos Seropédica, Vassouras and Piraí (RJ), and their influences on the yield of the bean crop (Phaseolus vulgaris L.). Cienc. e Agrotec. , Lavras, v.23, n.2. p. 323-330. 1999.

Neto, P. C.; Vilela, E, A. Veranico: a problem of drought in the rainy season. Climatologia Agrícola, Informe Agropecuário, v. 12, n. 138, p. 59-61. 1986

MACHADO, Paula. Fires cause concern in the north of Minas Gerais. 2009. Available at:<http://www.webcitation.org/query?url=http://www.gazetanortemineira.com.br/not i cias .php?id=3703+&date=2011-05-14>. Accessed on: 16 March 2016.

CHAPTER 3

EVALUATION OF THE EROSIVE POTENTIAL OF RAINFALL IN BELO HORIZONTE, MINAS GERAIS, FROM 1961 TO 2014

Daniel Dantas[1] ; Eduarda Gabriela Santos Cunha[2] ; Gabriela Paranhos Barbosa[3] ; Maria José Hatem de Souza[4] ;

[1]Forestry Eng., Master's Student, UFLA, Lavras-MG Phone: (38)99123-7493, dantasdaniel12@yahoo.com.br;[2] Forestry Eng., Master's Student, Forestry Eng. Dept., UFVJM, Diamantina-MG;[3] Forestry Eng. Forestry Engineer, PhD Student, Forestry Engineering Department, UFVJM, Diamantina-MG;[4] Agricultural Engineer, Associate Professor, Agronomy Department,
UFVJM, Diamantina-MG.

ABSTRACT: The aim of this study was to estimate the average monthly (EI30) and annual (R factor) rainfall erosivity and compare the variation in these factors between 1961 and 2014 in the municipality of Belo Horizonte, Minas Gerais. Monthly rainfall data from historical series of the National Meteorological Institute (INMET) and the National Water Agency (ANA) were used. To assess the variation in erosivity over time, the historical series used was divided into five ten-year periods and the average monthly rainfall for each period was calculated. The average annual rainfall erosivity was obtained by adding up the monthly EI30 values. The values found ranged from 8141 MJ.mm/ha.h.year to 9457 MJ.mm/ha.h.year. Despite the difference of 1316 MJ.mm/ha.h.year between the periods with the highest and lowest R factors, all periods fell into the high erosivity class. EI30 was closely related to the volume of precipitation. The rainy season was responsible for approximately 96 per cent of erosivity, while the dry season contributed 4 per cent.

KEY WORDS: erosivity; precipitation; water erosion.

INTRODUCTION

Erosion is the main form of soil degradation. In the erosion process, among the various factors related to the occurrence of erosion, rainfall erosivity is one of the most important (Mello et al., 2007).

Knowing the erosive potential of rainfall in a region is an important factor for controlling it in each region, due to its high temporal and spatial variability. Knowledge of erosivity, both in space and time, is fundamental for planning soil and water management and conservation practices aimed at minimising the adverse effects of water erosion (Viola et al., 2014).

Soil degradation is mainly caused by the dragging away of smaller, more nutrient-rich particles, culminating in a decrease in fertility and, consequently, a reduction in yields or an increased need for fertilisers and correctives. Soil losses caused by water erosion reduce the thickness of the soil, decreasing the capacity to retain and redistribute water in the profile, which leads to greater surface runoff and sometimes higher rates of soil erosion (Santos et al., 2010). In short, water erosion has a high potential for reducing the productive capacity of soils and can jeopardise surface water resources (Mello et al., 2007).

According to Santos et al. (2010), the association of features such as a steep terrain, inadequate land use and management, and adverse physical and water characteristics with more intense and frequent rainfall increases the risk of erosion.

Rainfall erosivity, known as the R factor, is part of the universal soil loss equation and is a numerical index that expresses the capacity of rainfall expected in a given location to cause water erosion in an unprotected area (Bertoni; Lombardi Neto, 1993). It can be expressed using indices based on the physical characteristics of rainfall in each region (Cabral et al., 2005).

To determine the erosive potential of rainfall, indices are used, such as the standard erosivity index, called EI30, described by Wischmeier and Smith (1978), which represents the product of the kinetic energy with which the raindrop hits the ground and its maximum intensity. According to Bertoni and Lombardi Neto (1993), this product represents an interaction term that measures the effect of how erosion by impact, splash and turbulence combines with the downpour to transport the detached soil particles.

Determining erosivity values throughout the year makes it possible to identify the months in which the risk of soil and water loss is highest - which is why it plays an important role in planning conservation practices based on maximum soil cover in the critical seasons when rainfall has the greatest erosive capacity (Wischmeier; Smith, 1978; Bertoni; Lombardi Neto, 1993; Hudson, 1995). For this reason, many studies have been carried out in order to define the spatial distribution of rainfall erosivity in regions of Minas Gerais, as can be seen in the work of Aquino (2005), for the southern region of Minas Gerais; and Mello et al. (2007), for the state of Minas Gerais.

In view of the above, the aim of this study was to estimate the average monthly (EI30) and annual (R factor) erosivity and compare the variation of these factors in the time series from 1961 to 2014 for the municipality of Belo Horizonte (MG). According to Lucio et al. (1998), due to its geographical location, Belo Horizonte (MG) is influenced by meteorological phenomena at medium and tropical latitudes and has a marked topography where there is an altimetric variation, making it a favourable area for water erosion.

MATERIALS AND METHODS

The municipality of Belo Horizonte is located in the south-eastern region of Brazil, in the centre of the state of Minas Gerais (Figure 1). It is located in the São Francisco River Basin, **at the following coordinates: 19°55' S and 43°56' W, at an altitude of 633 metres above** sea level. According to Koppen, Belo Horizonte's climate is classified as mesothermal Cwa. However, according to Assis (2012), when analysing the parameters for Koppen's climate classification, it can be seen that they do not have many adaptations to regional variables, especially in relation to topographical characteristics, a situation found in the municipality of Belo Horizonte. Assis **(2012) therefore proposes a classification into "natural" climatic units, subdividing the** municipality of Belo Horizonte into the Tropical High Altitude Climate of the Belo Horizonte Depression and the Tropical High Altitude Climate of the Iron Quadrangle Mountains. These climatic units are subdivided into their respective mesoclimates and topoclimates.

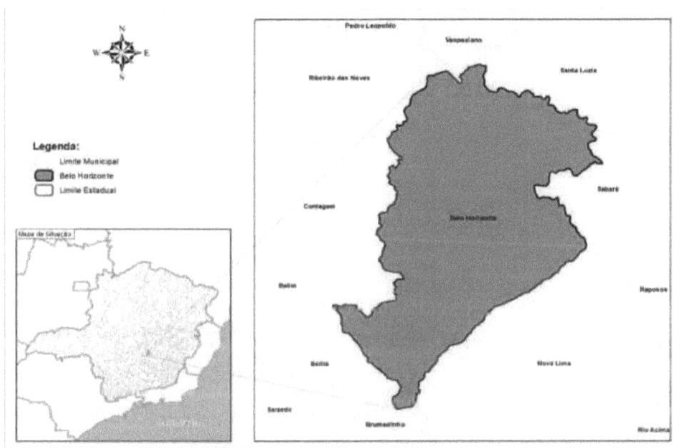

Figure 1 - Map showing the location of the study area.

Monthly rainfall data from historical rainfall series obtained from the 5th Meteorological District - 5° DISME - belonging to the National Meteorological Institute (INMET) and the Hydrological Information System of the National Water Agency (ANA) were used. The INMET meteorological station in Belo Horizonte is located at a latitude of 19.93° S, a longitude of 43.93° W and an altitude of 915.0 metres.

The data used in this study refers to the years 1961 to 2014. Five periods were defined for comparison: period 1 (1961 to 1971), period 2 (1972 to 1982), period 3 (1983 to 1993), period 4 (1994 to 2004) and period 5 (2005 to 2014).

To determine the average monthly erosivity index (EI30), the monthly average for each of the five periods evaluated was calculated using equation 1, proposed by Wischmeier and Smith (1978):

$$EI_{30} = 67{,}355 \left(\frac{r^2}{P}\right)^{0{,}85} \quad \text{(Eq. 1)} \qquad (1)$$

in which:

r: average monthly rainfall (mm)

P: average annual rainfall (mm)

The annual rainfall erosivity index (R) is the sum of the monthly values of this index, according to equation 2:

$$R = \sum_{1}^{12} EI_{30} \quad \text{(Eq. 2)}$$

The values found for the EI30 and the R factor were compared in order to check for changes in the time series studied. The classes of interpretation of the mean annual erosivity index (R) from the classification by Foster et al. (1981) Table 1 were considered for comparison with the values found for the R factor.

Table 1 - Interpretation classes of the mean annual erosivity index (R)

Erosivity (MJ.mm/ha.h.year)	**Erosivity classes**
R≤2452	Low
2452 < R ≤ 4905	Average
4905 < R ≤7357	Moderate
7357<R≤810	High
R > 9810	Very high

Source: (Foster et al., 1981).

RESULTS AND DISCUSSION

The average annual erosivity values (R Factor) for the periods under study are shown in Table 2.

Table 2 - Average monthly rainfall and R factor for each period evaluated

Periods	Average annual rainfall	R Factor
	mm	MJ.mm/ha.h.year
1961-1971	1440,4	8658,6
1972-1982	1445,1	8140,9
1983-1993	1603,5	9314,0
1994-2004	1592,5	9457,0
2005-2014	1610,5	9299,1

The values found for average annual erosivity (R Factor) were high, placing the study area in the high erosivity class, as the results are between 7357 and 9810 MJ.mm/ha.h.year, according to the classification by Foster et al. (1981), shown in Table 2.

When comparing the periods with the highest (period 5: 2005-2014) and lowest (period 2: 1972-1982) R factors, there is a difference of 1,316 MJ.mm/ha.h.year.

The dynamics of average monthly rainfall (mm) and the average monthly erosivity index (MJ.mm/ha.h) over the entire time series studied are shown in Figure 2.

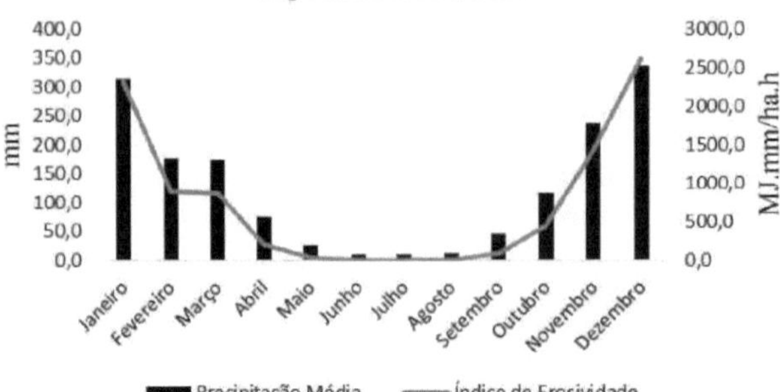

Figure 2 - Behaviour of average monthly rainfall (mm) and average monthly erosivity index (MJ.mm/ha.h) over the entire time series studied.

It is clear that the average annual erosivity index is closely related to the volume of precipitation. It can be seen that over the years there has been an increase in average annual rainfall. This increase was accompanied by the values of the R factor, except in period 4 (1994 to 2004) when there was a drop in average annual rainfall compared to the previous year, however, the R factor reached its maximum value in this same period. This may have been because period 4 (1994 to 2004) had the highest average rainfall in the rainy season, with 237.3 mm, meaning that heavy rainfall was concentrated in this period.

The study region has a well-defined dry and rainy season in which the driest quarter is June, July and August and the quarter with the highest rainfall is November to January. The months of January and December stand out, as they had the highest average rainfall values over the entire time series evaluated, with 314.4 and 336.7 mm, respectively. These rainfall figures are higher than the average rainfall for the rainy season, which is 226.1 mm.

According to Mello and Silva (2009), regions at higher latitudes and altitudes, where the Cwb/Cwa type of climate normally predominates according to the Koppen classification, have higher precipitation totals.

The average erosivity index for January and December also showed high values of 2333.5 MJ.mm/ha.h and 2616.2 MJ.mm/ha.h, respectively.

It was observed that the rainy season was responsible for approximately 96 per cent of erosivity, while the dry season contributed 4 per cent.

CONCLUSION

The months with the highest average rainfall were January and December, and consequently these were also the months with the highest average erosivity index values.

Rainfall and erosivity are closely related, and the increase in average annual rainfall was accompanied by an increase in the average annual erosivity index (R Factor).

ACKNOWLEDGEMENTS

To the Minas Gerais Research Foundation for financial support.

BIBLIOGRAPHICAL REFERENCES

Assis, W. L. 2012. The natural climates of the municipality of Belo Horizonte - MG. ACTA Geográfica, Boa Vista, Ed. Esp. Climatologia Geográfica 1: 115-135. Doi: 10.5654/actageo2012.0002.0008

Aquino, R. F. 2005. Rainfall patterns and spatial variability of erosivity for the south of the state of Minas Gerais. Lavras, MG: UFLA. 98 f. Dissertation - Federal University of Lavras.

Bertoni, J.; Lombardi Neto, F. 1993. Soil conservation. Ícone, São Paulo, BR.

Cabral, J. B. P.; Becegato, V. A.; Scopel, I.; Lopes, R. M. 2005. Study of erosivity and spatialisation of data with geoprocessing techniques on the topographic map of

Morrinhos-Goiás/Brazil for the period 1971 to 2000. GeoFocus 5: 1-18.

Foster, G. R.; Mccool, D. K.; Renard, K. G.; Moldenhauer, W. C. 1981. Conservation of the universal soil loss equation to SI metric units. Journal of Soil and Water Conservation 36: 355-359.

Hudson, N. W. 1995. Soil conservation. 3. ed. Batsford, London, UK.

Lucio, P. S.; Abreu, M. L.; Toscano, E. M. M. 1998. Characterisation of climatological series in Belo Horizonte - MG. PART I, II,III, IV. Anais... X Brazilian Congress of Meteorology. Brasília, DF, BR.

Mello, C. R.; Sá, M. A. C.; Curi, N.; Mello, J. M.; Viola, M. R. 2007. Monthly and annual rainfall erosivity in the state of Minas Gerais. Pesquisa Agropecuária Brasileira 42(4): 537545.

Mello, C. R.; de Silva, A. M. 2009. Statistical modelling of monthly and annual precipitation and the dry season for the state of Minas Gerais. Brazilian Journal of Agricultural and Environmental Engineering 13(1): 68-74.

Santos, G. G.; Griebeler, N. P.; Oliveira, L. F. C. 2010. Intense rainfall related to water erosion. Brazilian Journal of Agricultural and Environmental Engineering 14(2): 155-123.

Wischmeier, W. H.; Smith, D. D. 1978. Predicting rainfall erosion losses: a guide to conservation planning. Washington, USDA, (Agriculture Hand-Book, 537).

CHAPTER 4

USING ARTIFICIAL NEURAL NETWORKS TO PREDICT RAINFALL DURING RAINY PERIODS

Daniel Dantas[1] ; Eduarda Gabriela Santos Cunha[2] ; Gabriela Paranhos Barbosa[3] ; Maria José Hatem de Souza[4] ;

[1]Forestry Eng., Master's Student, UFLA, Lavras-MG Phone: (38)99123-7493, dantasdaniel12@yahoo.com.br;[2] Forestry Eng., Master's Student, Forestry Eng. Dept., UFVJM, Diamantina-MG;[3] Forestry Eng. Forestry Engineer, PhD Student, Forestry Engineering Department, UFVJM, Diamantina-MG;[4] Agricultural Engineer, Associate Professor, Agronomy Department,
UFVJM, Diamantina-MG.

ABSTRACT: The aim was to estimate rainfall in the rainy season based on rainfall in previous dry seasons in Diamantina, using Artificial Neural Networks (ANN). The chronological order of the data was changed so that the dry season of one year was related to the rainy season of the following year. Part of the data was used for training and part for evaluating the ANN's performance. Time series analysis was used and the best network found was the radial basis function type. The ANN had an average error of 10 per cent. The average rainfall during the period in which the network was applied was 1099 mm, while the average estimate was 1128 mm. Using data from the dry periods to estimate rainfall in the rainy period gives satisfactory results and changing the chronological order of the dry period resulted in a more effective forecasting network than not changing it.

KEYWORDS: climate scenarios, climate modelling, weather forecasting.

INTRODUCTION

The agricultural development of a region is strongly related to its rainfall, with a shortage of rainfall being an unfavourable factor for the production process. Rainfall is the most economical and environmentally correct way of using water in agriculture, because as well as helping to maintain the balance of existing water resources in another region, the need to import water through irrigation considerably increases production costs.

According to Castro (1994) and Andrade et al. (2009), many Brazilian producers use the average monthly rainfall to dimension their agricultural projects. Climate modelling is therefore an important tool, as it allows future scenarios to be simulated, both for agricultural and industrial activities, as well as for prior knowledge of possible tragedies, studies and planning of water resources, decision-making in relation to electricity generation and other activities (QUEIROZ et al., 2001; SAMPAIO et al., 2007; ÁVILA et al., 2009). Climate forecasting of precipitation is an important issue in meteorology, since it is a variable associated with natural disasters (droughts and floods) and agricultural harvests, with impacts on the tourism and transport sectors (ANOCHI, 2015). However, this meteorological variable is difficult to predict due to its great temporal and spatial variability (discontinuous variable).

With the climate and its changes coming to the fore in the last decade, scientists around the world are trying to understand the nature of the changes that are likely to occur, as well as the possible impacts they could have on society in general (ANOCHI, 2015).

Various tools can be used in the meteorological field to estimate rainfall and, among them, Artificial Neural Networks (ANN) stand out, which have shown satisfactory results in climate projection (MOREIRA et al., 2006; ANOCHI et al., 2009). In simple terms, according to Oikawa and Ishiki (2013), Artificial Neural

Networks can be defined as a statistical tool whose operating principle is governed by a mathematical model inspired by the functioning of the basic elements that make up the neural structure of intelligent organisms, which acquire knowledge through experience and, by processing information, generate an output (predicted data) from one or more inputs presented (predictors). ANNs are made up of neurons or processing units that compute certain mathematical functions, usually non-linear (ANOCHI, 2015). These processing neurons can be distributed in one or more layers and interconnected by a large number of connections (synaptic weights), which store the knowledge represented in the model and serve to weight the input received by each neuron in the network. Through successive presentations of previously known input and output data, the ANN learns the relationship between them (input and output) and, using an error minimisation algorithm, tries to reduce the mean square error in each training iteration.

Due to the physical complexity of precipitation processes, ANNs have great potential as a climate modelling and projection tool, since they have been successfully applied to a wide variety of problems, consolidating themselves as a technique for solving complex problems in pattern recognition, pattern classification, control systems, function approximation and predictive modelling (ANOCHI, 2015),

Sousa and Sousa (2010) proposed a model based on neural networks to simulate and predict average monthly flows at the fluviometric station located in the city of Piancó, in the semi-arid region of Paraíba. Ruivo et al. (2015) used data mining methodologies to investigate the causes of extreme weather events: the great droughts in Amazonas in 2005 and 2010, and the extreme rainfall in Santa Catarina in 2008, in which it was possible to pinpoint some of the climatological parameters responsible for these events.

The aim of this study was to evaluate the performance of Artificial Neural

Networks (ANN) for estimating rainfall in the rainy season, based on rainfall data from previous dry and rainy seasons in Diamantina-MG.

MATERIALS AND METHODS

The work was carried out at the Federal University of the Jequitinhonha and Mucuri Valleys (UFVJM), 6 km from the town of Diamantina - MG, located in the **Southern Espinhaço region, at an altitude of 1,149 m, 18° 17'S latitude and 43°34'W** longitude. According to the climate classification drawn up by Nimer (1989) for Diamantina, the climate is tropical with a subsequent climatic domain and a semi-humid subdomain, with a climatic variety of 4 to 5 dry months (IBGE, 1977). The predominant vegetation formations in the region are Campo rupestre and Cerrado rupestre. Daily rainfall data for the period 1977 to 2014 was obtained from the climatological station of the National Meteorological Institute - INMET, located at latitude 18.25°S, longitude 43.60°W and altitude 1296.9m. Data provided by the National Water Agency (ANA), located at latitude 17.61°S, longitude 43.60°W (Figure 1) and altitude 1300 m, was also used to complete some missing data from INMET's historical series for the months of February and March 1979 and July 1990. The data referred to daily rainfall from 9am one day to 9am the next, in accordance with the international standardisation of the WMO (World Meteorological Organisation), i.e. the amount of rainfall on a day (n) occurred between 9am the previous day (n-1) and 9am that day.

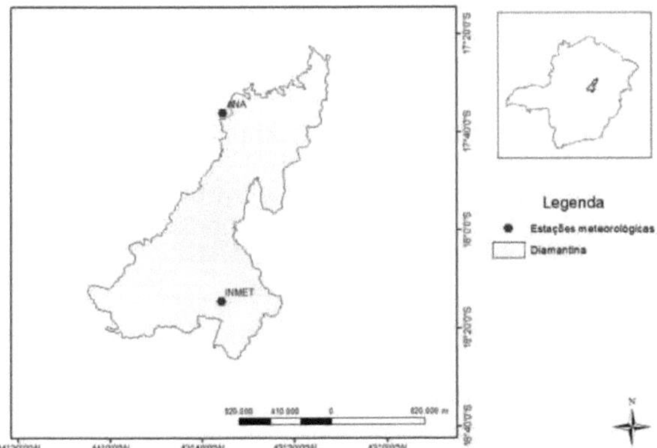

Figura 1. Location of the municipality of Diamantina, in Minas Gerais, and the meteorological stations of the National Institute of Meteorology - INMET and the National Water Agency - ANA, where the data was collected.

Vieira et al. (2010), in a study on the behaviour of rainfall during the rainy season in the region of Diamantina, MG, observed that the rainy season is between the months of October and March, and represents 88% of the total annual rainfall. Therefore, the data was divided into two periods: the dry period, corresponding to the sum of the rainfall between April and September, and the rainy period, corresponding to the sum of the rainfall between October and March.

The Artificial Neural Networks (ANN) were trained in Statistica 10 (Statsoft, 2014) using the Automated Neural Networks (ANN) tool, in which the training set consists of input and output pairs (x,yd), with the desired output yd being characterised beforehand for a given input value x. Adjusting the weights modifies the output y so that the difference between y and yd, i.e. the error, is reduced with each interaction (Oikawa and Ishiki, 2013). Time series analysis was used, which consists of a network trained with values from a time series that occurred over a given time interval, with the output being a future value from the series. All the networks used were of the Radial basis function type with 1 input layer, where the patterns are presented to the network; 1 hidden layer, which works as a feature recogniser that is stored in the synaptic weights and is responsible for most of the processing; and 1 output layer, where the network's output signals are presented. The number of neurons in the network's hidden layer was chosen automatically by the ANN tool.

The ANNs were trained in two ways: rainfall data from the dry season was used as an input variable to predict rainfall in the rainy season of the same year (situation 1); and rainfall data from the dry season was used to predict rainfall in the rainy season of the following year (situation 2). To do this, the chronological order of the rainfall data was changed so that the dry period of one year (x) was related to the rainy period of the following year (x+1). For example, the dry period from April to September 2010 was related to the rainy period from October 2011 to March 2012.

Rainfall data from the years 2010 to 2014 was not included in the training of the networks, as it was used exclusively to validate the chosen network.

Twelve training sessions were carried out for each group. In each session, 50 ANNs were trained and the network with the best performance in the training, test and validation values was retained.

The results obtained with the developed ANNs were evaluated based on the calculation of Average Relative Errors, ERM% (1) (Schaeffer, 1980), histograms with observed and estimated values and residue graphs.

$$ERM\,\% = \frac{(Vest - Vobs)}{Vobs} \times 100 \quad (1)$$

where: Vest is the estimated value and Vobs is the observed value.

RESULTS AND DISCUSSION

The average rainfall during the rainy season, considering all the years under study, was 1170 mm. With a maximum of 1632 mm in 1991 and a minimum of 445 mm in 1982. For the dry period, the average rainfall was 159.1 mm. After training the ANNs, the prediction of the rainy periods generated an average estimate of 1145 mm in situation 1, in which rainfall data from the dry period was used to predict rainfall in the rainy period of the same year. In this case, the Average Relative Error (ERM%) was 15.99 %. A maximum ERM% of 33.19 % was found in 2014, when the network estimated rainfall of 1,345.6 mm and the observed value was 899 mm (Figure 2). The minimum ERM% was 0.86 % in 1997, with estimated and observed rainfall values of 1093.6 mm and 1103.0 mm, respectively (Figure 2).

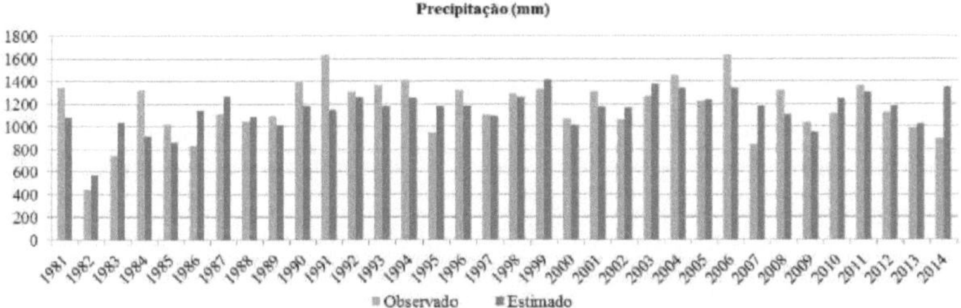

Figura 2. Precipitation observed and estimated by ANN using precipitation data from the dry season to predict precipitation in the rainy season of the same year.

In situation 2, in which the networks were trained using dry season rainfall data to predict rainfall in the following year's rainy season, the ANN showed an average estimate of 1160.7 mm, very close to the real figure (1170 mm), and an ERM% of 13.36 %. The maximum ERM% was 54.6 % in 1982, when there was a large variation in the amount of rainfall (445 mm) and, despite the high error obtained, the network managed to keep up with this variation and presented a low estimate of rainfall for the period (688 mm) (Figure 3). The minimum ERM% was 0.56 % in 1992, with estimated and observed rainfall values of 1307 mm and 1299.6 mm, respectively (Figure 3).

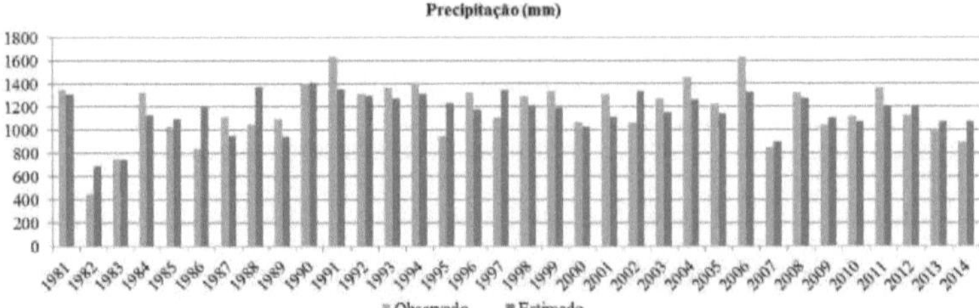

Figura 3. Rainfall observed and estimated by ANN using dry season rainfall data to predict rainfall for the following year's rainy season.

The relationship between observed and estimated rainfall in situation 1 showed a linear correlation coefficient of r = 0.67, indicating a moderate correlation, while in situation 2 there was a strong correlation, with a linear correlation coefficient of r = 0.81.

This shows that the ANNs produced satisfactory results and were able to describe the variations in rainfall over the period studied, using only data from previous dry periods as an input variable. However, there was a slight tendency for these variations to be smoothed out, since there was a greater ERM in very dry or very rainy periods.

Considering the data from the period used only to apply the ANNs (from 2010 to 2014), the average rainfall in the rainy season was 1099 mm, while the networks showed average estimates for the same period of 1221 mm in situation 1 and 1128 mm in situation 2, which again showed an average value very close to the real one. The ANNs showed an average percentage error of 11 and 10 per cent for situations 1 and 2, respectively

The use of ANNs as meteorological forecasting models has been applied in several studies, showing satisfactory results, including Srivastava et al., (2010); Talei et al. (2010) and Wu et al. (2010), showing their potential in this area of study. Zhang et al. (1997) and Menezes et al. (2015), who used artificial neural networks to predict rainfall, also obtained estimates with an ERM of 10%. The maximum ERM% was 33 % for situation 1 (Figure 4) and 19 % for situation 2 (Figure 5), while the minimum

ERM% was 3.5 % for both situations (Figures 4 and 5). This shows that the change in the chronological order of the data carried out in this study resulted in a slight improvement in the performance of the trained networks.

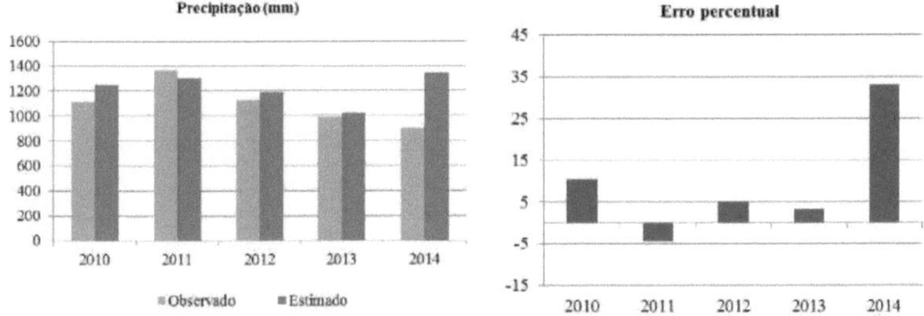

Figura 4. Rainfall observed and estimated by ANN using dry season rainfall data to predict rainfall in the rainy season of the same year. And percentage estimation error.

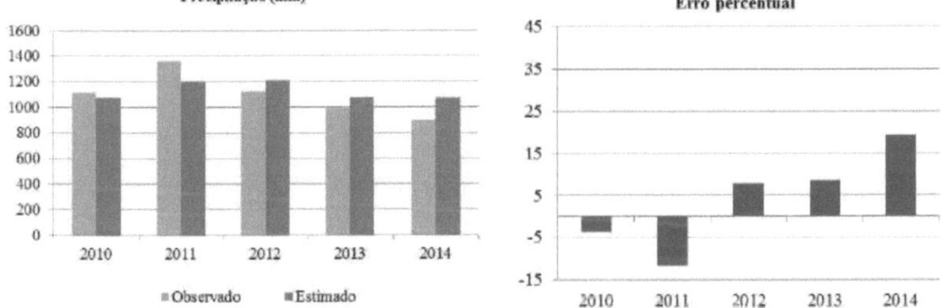

Figura 5. observed and RNA-estimated rainfall using dry season rainfall data to predict rainfall for the following year's rainy season. And percentage estimation error.

In summary, it can be said that the problem of predicting rainfall is not a simple task to model mathematically, since the variables that cause the phenomenon are highly non-linear and show complex behaviour. And meteorological variations, especially rainfall over a short period of time over the years, have no definite trend. However, it is possible to verify the satisfactory performance of the trained ANNs in estimating the rainy season, given that various factors influence the dynamics of rainfall over the years and, even so, the ANN showed an ERM% of 10% in estimating rainy seasons with only the information on rainfall during the dry season in previous years as the training input variable. By including more variables, it is possible to obtain even more accurate estimates. As such, ANNs have the potential to become an important tool for estimating climatic variables.

CONCLUSIONS

The Artificial Neural Networks manage to describe the variations in rainfall over the period studied, but with a slight tendency to smooth out these variations. The use of data from the dry and rainy periods to estimate rainfall in the rainy period shows satisfactory results and the change in the chronological order of the dry period resulted in a network with a more effective forecast than the non-change.

ACKNOWLEDGEMENT

To the Minas Gerais Research Foundation - FAPEMIG, for its financial support.

BIBLIOGRAPHICAL REFERENCES

ANDRADE, A. R. S.; FREITAS, J. C.; BRITO, J. I B.; GUERRA, H. O. C.; XAVIER, J. F. Application of conditional probability and the Markov chain process in analysing the occurrence of dry and rainy periods for the municipality of Garanhuns, PE, Brazil. Ambiente e água, Taubaté, v. 4, n. 1, p. 169-182, 2009.

ANOCHI, J. A.; SILVA, J. D. S. Use of artificial neural networks and rough set theory in the study of seasonal climate patterns. Revista da Sociedade Brasileira de Redes Neurais, v. 7, n. 2, p. 83-91, 2009.

ANOCHI, J. A. Climate forecasting of precipitation by autoconfigured neural networks. Thesis (Doctorate in Applied Computing), National Institute for Space Research, São José dos Campos, 2015.

ÁVILA, L.F.; MELLO, C.R.; VIOLA, M. R. Mapping the minimum probable rainfall for southern Minas Gerais. Revista Brasileira de Engenharia Agrícola e Ambiental, Campina Grande, v. 13, p. 906-915, 2009.

CASTRO, R. Probabilistic distribution of rainfall frequency in the Botucatu-SP region. Dissertation (Master's Degree in Agronomy), Universidade Estadual Paulista, Botucatu-SP, 1994.

BRAZILIAN INSTITUTE OF GEOGRAPHY AND STATISTICS - IBGE. Geography of Brazil: Southeast Region. Rio de Janeiro: IBGE, 1977.

MENEZES, P. L.; AZEVEDO, C. A. V.; EYNG, E.; DANTAS NETO, J.; LIMA, V. L. A. Artificial neural network model for simulation of water distribution in sprinkle irrigation. Brazilian Journal of Agricultural and Environmental Engineering, Campina Grande, v. 19, n. 9, p. 817822, 2015.

MOREIRA, M. C.; CECILIO, R. A.; PINTO, F. A. C.; PRUSKI, F. F. Development and analysis of an artificial neural network to estimate rainfall erosivity for the state of São Paulo. Revista Brasileira de Ciência do Solo, Viçosa, v. 30, p. 1069-1076, 2006.

NIMER, E. Climatology of Brazil. Rio de Janeiro: IBGE, 1989. 421p.

OIKAWA, R. T.; ISHIKI, H. M. Statistical and artificial neural network models used in precipitation prediction. Electronic Journal Fórum Ambiental da Alta Paulista, São Paulo, v. 9, n. 8, p. 19-34, 2014.

QUEIROZ, E.F.; SILVA, R. J. B.; OLIVEIRA, M. C. N. Periodic regression analysis model of monthly precipitation in the southeast Atlantic basin of Paraná. Pesquisa Agropecuária Brasileira, Brasília, v. 36, n. 5, p. 727-742, 2001.

RUIVO, H.; CAMPOS VELHO, H.; SAMPAIO, G.; RAMOS, F. Analysis of extreme precipitation events using a novel data mining approach. American Journal of Environmental Engineering, v. 5, n. 1A, p. 96-105, 2015.

SAMPAIO, S.C.; LONGO, A. J.; QUEIROZ, M. M. F.; GOMES, B. M.; BOAS, M. A. V.; SUSZEK, M. Estimation and distribution of decadal precipitation for the state of Paraná. Irriga, Botucatu, v. 12, n. 1, p. 38-53, 2007.

SCHAEFFER, D.L. A model evaluation methodology applicable to environmental assessment models. Ecological Modelling, Tennessee, v. 8, p. 275-295, 1980.

SOUSA, W.; SOUSA, F. Artificial neural network applied to flow forecasting in the Piancó River basin. Revista Brasileira de Engenharia Agrícola e Ambiental, v. 14, n.

2, p. 173-180, 2010.

SRIVASTAVA, G.; PANDA, S. N.; MONDAL, P.; LIU, J. Forecasting of rainfall using ocean-atmospheric indices with a fuzzy neural technique. Journal of Hydrology, v. 395, n. 3-4, p. 190-198, 2010.

TALEI, A.; CHUA, L. H. C.; QUEK, C. A Novel Application of a Neuro-Fuzzy Computational Technique in Event-Based Rainfall-Runoff Modelling. Expert Systems with Applications, v. 37, n. 12, p. 7456-7468, 2010.

WU, C. L.; CHAU, K. W.; FAN, C. Prediction of Rainfall Time Series Using Modular Artificial Neural Networks Coupled with Data-Preprocessing Techniques. Journal of Hydrology, v. 389, n. 1-2, p. 146-167, 2010.

ZHANG, M.; FULCHER, J. SCOFIELD, R. A. Rainfall estimation using artificial neural network group. Neurocomputing, v. 16, n. 2, p. 97-115, 1997.

Printed by Books on Demand GmbH, Norderstedt / Germany